The Author

David Kaye started his career in market research and then worked on planning models for Shell International for some years in London and Paris.

He joined Arthur Andersen & Co. in London in 1962 when there were only 30 people in management and information consulting. Within five years he rose to partner, helping to build up what has become Andersen Consulting with 1000 people in the UK. He was responsible for commercial and public sector consulting. More recently he has been concentrating on the dependence of strategy and organisation on information technology.

Married with three children, David Kaye was educated in mathematics at Cambridge, in statistics at Oxford and in operational research at Ann Arbor in Michigan USA.

To my colleagues and clients, from whom I have learnt
most of what I know.

GAMECHANGE

**The impact of information
technology on corporate
strategies and structures**

**An Andersen Consulting
Management Briefing**

A boardroom agenda by David Kaye

Heinemann Professional Publishing

"The history of IT can be characterised as the overestimation of what can be accomplished immediately and the underestimation of long-term consequences"

Paul Strassman *Information Payoff*

Heinemann Professional Publishing Ltd
Halley Court, Jordan Hill, Oxford OX2 8EJ

OXFORD LONDON MELBOURNE AUCKLAND SINGAPORE
IBADAN NAIROBI GABORONE KINGSTON

First published 1989
© Andersen Consulting 1989

British Library Cataloguing in Publication Date
Kaye, David
  Gamechange
  1. Management. Decision making. Applications of computer systems
  I. Title
  658.4'03'0285

ISBN 0 434 91022 8

PREFACE

This book is intended for chief executives and their colleagues who are thinking about how the Information Technology revolution will affect the strategies and structures of their businesses. Management's task is difficult enough without having to penetrate "computerspeak", so I have avoided it where I can. But where I cannot, I have tried to make its meaning plain.

I have selected the few key facts about the advances in IT that have to be understood and have tried to explain where I think they lead. I could not have done this without reference to the experience of companies who are leading the way in changing their strategy and structure. Above all I could not have attempted the task without the support and guidance of my Andersen Consulting colleagues. While there are many who have helped with examples and argument, I want above all to recognise the help from my partners John Hollis, Keith Ruddle, Bill Lattimer and Bob Dymond, and of Paul Thorley who did much of the work sorting out what we think matters. But of course, as the author always says, and this time really means, the errors and omissions are mine. And for patient organisation of the text from draft to draft to draft, I thank my unflappable secretary, Sue Rider.

David Kaye

David Kaye is the partner in charge of strategic consulting in Andersen Consulting, London.

Contents

Introduction

AIMS AND APPROACH

The background

It is obvious that we are in a period of revolution in information technology. The results of the thirty year transformation in capability and costs of computers and telecommunications are everywhere to be seen, whether in financial market institutions, computer controlled production lines, or the home. This transformation has been powered by advances in basic science and technology. These advances will continue at a similar pace for many years yet. Therefore the information technology (IT) transformation will continue. And, as it continues, it will broaden and deepen its effects on the information infrastructure of enterprises.

The condition of the information infrastructure is key to the strategy of an enterprise. And it is key to the management arrangements which are required to deliver the strategy. The Romans knew this when they built their roads. To expand and maintain their frontiers in the face of the enemy, they built communication highways to support their relative advantage of discipline and wealth. And they built management structures that could respond effectively to the ways in which information flowed along the highways and in their societies. They used the best available information technology, and they defined the frontiers of national activity, and of individual responsibility, to fit with that technology.

It is no different for the enterprise of today, operating in a world where sustained competitive advantage is an aim of the development of strategy and structure. But the speed and depth of change in the available technology make it perhaps more difficult to adapt strategy and structure accordingly.

In one way or another, management has a general awareness of what has been happening and an unease with its consequences. Computer features in the press or in the science columns of journals such as *The Economist* set out what has been happening in the IT related fundamentals of science and technology. Exhortations to become directly involved with corporate issues of computing are loud and frequent. The costs of IT can be large enough to be significant at the corporate level and to be a factor in the chief executive's mind as he tries to fit corporate objectives within the constraints imposed by finance and the politics of the board and the stockmarket. The business benefits do not always follow. All of which may well start a worry in the chief executive's head along the lines of "Something is going on out there; it's important; leaving it to the IT people to fix doesn't seem to be good enough; but I can't do everything, and I didn't work with computers when I was young; the ground is shifting under me; the game is being changed, and I need to know how".

The aim of the book

This book addresses the three aspects of all this which really ought to be the concern of the chief executive and his top management team as a consequence of the impact of the IT revolution. First, what questions should be asked about strategic direction for the company and for its functions? Second, what questions should be asked about corporate structure and management arrangements? Third, how should the work of getting and implementing the answers be coordinated?

This book points the way in some important areas of management concern to questions about strategy and structure which follow from the technical developments, and which management should be asking. The purpose is not to point to developments in information systems as such, nor is it to discuss the many – and often more important – issues of strategy and structure which have little or nothing to do with IT. The purpose is rather to provide insights which should help management and their advisers ask the questions which have become worth asking as a result of IT developments, and find the answers.

What the book does not try to do is make observations to IT professionals about how IT can be made to do its job better; the reader will find nothing about advances in the planning, design, and delivery of new computer systems, nor

Systems drive, and are driven by, strategy and structure . . .

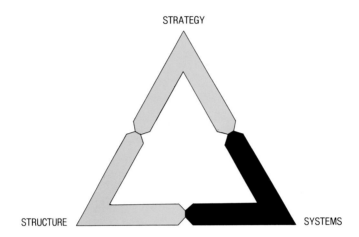

but only some of the effects on systems of science and technology affect strategy and structure

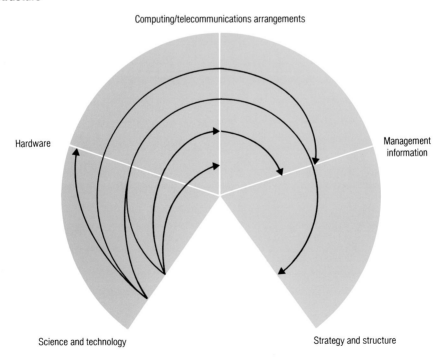

about controlling the costs of IT investment.

The IT revolution is built on some basic developments in science and technology. These have transformed computer and telecommunications hardware. In turn, the capabilities and costs of the new hardware have changed how the hardware should be best arranged, that is to say, where computers should be placed, under whose control, to do what, and how they should be linked via telecommunications. As a result, the characteristics of information which can be available for company use have been transformed.

Some, but by no means all, of these transformations change the assumptions and compromises which lie behind management's definition of strategy and structure. Some of these changes are so important that strategy and structure now need to be re-examined in the light of the IT revolution. It is the aim of this book to stimulate management to focus on those issues of business strategy and organisation affected in an important way by the IT revolution, and to support management in the re-examination task. This top management task can be critical to securing competitive advantage. It will sometimes be more important than the different challenge to management – and especially IT management – to continually reassess IT strategy in the light of developments in IT and of changes in business strategy and structure. The theme of the book is to help top managers work their way to a vision of a competitive future, exploiting the new power of IT, but tempered by restraints of practicality and economics.

The book is based on the belief that management will feel more comfortable in following through the technology changes and acting on them if they first appreciate the nature of those parts of the science and technology and consequential changes that are driving redefinition of the frontiers of strategy and structure. The first part of the book therefore aims to highlight the key features of IT developments along the paths which connect science and technology via hardware and system changes to corporate strategy and structure. This part is intended to prepare the ground for the subsequent main parts on strategy and structure.

The structure of the book

The book is organised in four parts.

The first part sets out briefly the essential facts and jargon associated with the IT revolution which top management need to appreciate so that they can focus on related key areas of strategic concern.

The second part draws attention to issues of strategic direction affected by IT. This part is divided into chapters on strategy for business unit boundaries, markets, products and services, production and distribution, and human resources.

The third part, divided into chapters on middle management and top management, sets out some important organisation structure and management issues affected by IT.

The fourth part outlines a framework for integrating the formulation of strategy for direction and structure with the planning of implementation of strategic change, and then concludes with a summary of the book's argument and view of management's next moves.

The IT revolution is a driver of corporate turbulence and competition. Those chief executives who do not ride with the revolution will probably lose. Those who ride with it have a chance of winning. This book is intended to help chief executives direct the ride and keep control.

PART ONE

A layman's guide to the key
facts that affect business
strategy and structure

The information revolution

1

SCIENCE AND TECHNOLOGY EFFECTS
ON HARDWARE

Computers are becoming much cheaper and smaller

Computer power has been made dramatically cheaper – 25 percent per annum over the last twenty years or so – and this trend will continue. Computers have also become smaller and more convenient to use; some have become portable; and access to a computer can sometimes now be through devices which can be held in the hand. These changes have been possible because of the continually increasing capability to "miniaturise" electric circuits on semiconductor chips.

Semiconductor chips are basic to the modern electronic computer

Chips have usually been made of silicon. This material conducts electricity in a manner which is part way between that of a conductor – (such as copper) – and an insulator – (such as glass). Hence a "semiconductor".

A chip of silicon can be very finely etched with chemicals. These chemicals can be chosen to change the electrical characteristics of a microscopic part of the chip's surface in various ways. This is what gives a chip its importance. It means that a chip – only a few millimetres in size – can be etched to carry large numbers of electronic components such as transistors, resistors, and capacitors, with connections between them to form whatever integrated circuits the computer designer requires.

The semiconductor chip has major advantages. It is made of very cheap material. It takes up very little space. It uses very little electricity and puts out very little heat. And assembling a circuit has to be done only once so that it can be photographed. The photograph is then etched onto all the chips made to the particular design, so that costs of one assembly are widely spread.

The greater the complexity of network that can be carried on a small chip, then the greater are the advantages of the chip.

Advances in technology are opening the way to chips which are more complex and cheaper

Lithographic techniques are making it possible to etch a hundred times more components on a chip than ten years ago.

The basic chip material – silicon – will sometimes be replaced by new materials (such as gallium) which have superior properties for some types of circuit. New materials linked with continued improvements in lithography will bring a further hundredfold increase in component density in the next ten years.

It is becoming economic to custom design a chip to carry out a particular business function rather than a general computer function, thus further reducing the cost of a computer and the cost and time of developing programs for a particular job.

The billion component chip is in prospect

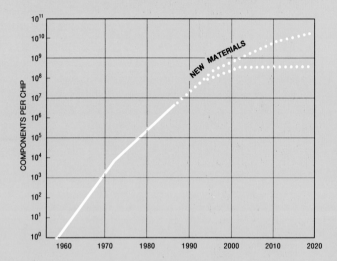

Telecommunications networks for the transmission of data are becoming more cost-effective

The technologies of digital transmission and fibre optics are transforming the effectiveness and economics of linking computers in communicating networks. They are also leading to the possibility of both voice and image being carried on the same networks.

Digital transmission is key to cost effective linking of computers in a telecommunications network

Information in a computer is present as a series of on/off pulses or digits. The ability to transmit information in digital form is therefore the key to linking computers so that information can be communicated electronically from one computer to another.

Digital transmission can also be used for communicating voice and images. The advantage is that not only can the same physical network of wires and switches be used as for linking computers, but also, as every hi-fi enthusiast knows, by converting a wave signal into a series of digital pulses, distortions of the signal in transmission are minimised. Also the capacity of a network can be used much more efficiently by information in digital form.

Fibre optic technology improves the economics of telecommunications networks

Digital transmissions, converted into pulses of light, can be sent along thin glass fibres. This optical communications technology improves a thousandfold the carrying capacity of a cable and should greatly improve the cost effectiveness of digital transmission. Optical technology offers the prospect of yet further cost improvement and of opportunity for an integrated system of public and private data "highways".

The cost effectiveness of storing data has been improving vastly

Technologies for magnetic storage of data in computers are greatly increasing computer storage capacities and reducing costs. As optical methods are developed, data storage will further improve, allowing combinations of sound, pictures, and computer data to be stored on a single disk.

Major advances in the technologies of compact storage of accessible data are key

Disks covered with a magnetic surface are used to store electronic data. Just as with semiconductor chips, technology has been increasing the density with which data can be stored: from 30,000 characters per square centimetre in 1968 to more like 3 million characters by the early 1990s. This reduces costs and speeds up the process of searching for data in very large blocks of storage.

As with hi-fi compact discs, optical methods using laser monitoring of data stored on cheap and light plastic discs are opening up fast access to very large amounts of data.

Cheap storage and computing power are becoming available in pocket sized cards

The so-called "smart card", looking like a credit card but incorporating a chip, can be used like a credit card but with the extra feature that data in magnetic storage on the card can be updated when a transaction is made. Laser cards are similar but, using laser technology rather than magnetic technology, can carry vastly greater data storage. These cards provide a very simple way for a user to link in with a computer system.

2

HARDWARE EFFECTS ON SYSTEMS

People needing to process information can now have computing under their direct control

Miniaturisation changes the relative costs, convenience and capabilities of a central computer serving several users in favour of several smaller computers each serving one group of users: the "distributed processing" which can result is done by appropriately sized minicomputers and microcomputers, often connected in a "network" – and with a central computer – to exchange data and information within an office or between offices across the country and the world.

As optical communications technology improves, so will telecommunications costs fall, making distributed processing even more cost effective in those situations where it is appropriate.

Links can be established between computers in different companies

Not only can links be established between computers in different offices, but the offices can be in different companies using different computer systems. Companies can do this by subscribing to a common network service which is able to convert a company's data into forms accessible to other subscribers. Such a service is an example of a so-called Value Added and Data Service (VADS).

Optical communications and improvements in programs for operating networks will provide continued impetus for more network services; these will replace many telephone and written communications and make for much closer and detailed connections between different parts of a supply – manufacturing – distribution chain.

The vast amount of corporate detailed data can now be an electronically accessible resource

The changes in the technology of data storage have led to it being practical to store very large amounts of data in computer readable form, in what has become commonly known as a database.

A database can be accessible to many users, each of whom does not therefore have to be responsible for putting all the data he may want to use into the computer – thus avoiding duplication of clerical effort and confusion between inconsistent versions of the same data.

For most efficient use of the storage capacity, the positioning of the different items of data in the database assumes the data is used in fixed combinations. Users of a database organised in such a manner have to obey the rules of the system and are constrained in how they can use the data. This can be frustrating. Also it can be expensive and can take a long while to change the arrangement of the data when new requirements are placed on it (for example, after a reorganisation, or regrouping of products).

As storage costs continue to decline, the importance of the efficiency of storage declines. This means that the database designer's priority can shift from efficiency to flexibility. In the new designs, data is accessed via the relationships that exist between the different types of data. The resulting "relational database" goes some way to remove the constraints over the user in how he gets data combined and reported. This introduces the flexibility, which is essential for rapid adaptation to new user requirements.

The distributed database is on its way

Distributed processing with a consistent database available to all users is now beginning to be a reality as new software becomes available to manage the various database and network operations. The availability of powerful "network management" tools means that a user will be able to work with and change information which can then be immediately available to a user in another part of the office or the world.

Very complex operations can be supported with the new database technology

The new low-cost, high-capacity, high-flexibility databases are making it possible to handle the vast amounts of information which may be required for operations such as designing a new engineering component. This has been key to developments such as establishing vast engineering databases and linking them in systems that include facilities such as for computer aided design (CAD).

3

SOFTWARE DEVELOPMENTS

Advances in methods of preparing computer programs have been improving the costs and benefits of computer systems

Until the mid 1980s about half the cost of setting up a computer system was absorbed in programming software – that is, writing the instructions for the computer to operate. About half of the programming effort can now be saved through the use of various software building tools and blocks of standard software which can be made available on the programmer's computer. The opportunity for this improvement in programmer productivity provides a significant offset to the increasing costs of programmers.

Often at least as important is the effect of new methods for developing systems. These "systems development methodologies and tools" can reduce the elapsed time for development of a working system, from years to months, and months to weeks, producing a major change in the ability of systems designers to respond to new management requirements.

Software developments are making computers easier to use

Computers are becoming easier for people to use – computers are becoming "friendly" to ordinary people. Computers are now being seen as part of the corporate scenery; barriers to their acceptance by staff are falling. The change in attitude has been helped by designers concentrating on the "interface" between users and computer. They have built software into the computer system which allows the user to "converse" with the computer, exploiting the senses of sight, touch and hearing. And so-called "expert systems" are beginning to capture some of the processes of thinking in particular areas of expertise.

INFORMATION DEVELOPMENTS

Five developments are key to strategy and organisation

Use of the new technology has brought about five information developments which can have a profound effect on what are the best strategies and organisation for a business. *The test for whether strategies or organisation should be questioned in the light of the IT revolution is the importance to the business of these five information related improvements.*

Two improvements make information easier to use and with less work . . .

- *Time economy:* the new technology can reduce the staff time required to record, reconcile, retrieve, analyse and present data.
- *Accessibility:* relatively easy access can be arranged to data which is consistent and up-to-date, whether or not the data is held in different departments or locations.

and three improvements make information more useful. . . .

- *Timeliness:* the most modern developments reduce the delay with which information can be made available.
- *Computability:* calculations can be made quickly, accurately, and with little staff effort to turn data into the forms of information which are most useful for management analysis – whatever complexity of calculations is required.
- *System adaptability:* the collection and presentation of information can be more readily adapted to fit changes in strategy and organisation: new systems development methodologies and tools allow changes to be made in the necessary systems with less delay; hardware and systems are becoming more "friendly" and responsive to user requirements; and some elements of cost are falling.

As major technologies develop, the costs of an information system fall and the user convenience elements of effectiveness rise . . .

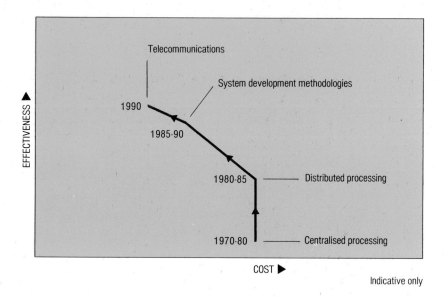

Indicative only

but it is now cost effective to seek more from using IT. So total IT costs are likely to rise

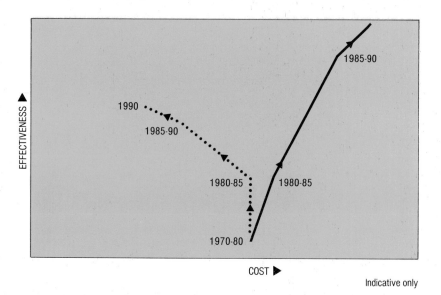

Indicative only

The technology gap is shrinking

Inevitably there is a gap between, on the one hand, the amount and quality of information which the technology can make available within an organisation and, on the other, the information which would ideally be most useful. But the trend of information developments is that this gap is shrinking. The trend is towards giving an authorised member of a company's staff the ability to get hold of the data about the past and the up-to-date data about current operations which he needs to do his job, in the form he wants, wherever the data may have been placed in the company's computerised system – whether local or global – and analyse it as he wishes.

Furthermore, and importantly, the trend is towards this data accessibility and computability being achieved easily and economically.

How far this trend in a company and in an industry – has affected the jobs that can be done and that are worth doing can have a profound effect on many issues of strategy and structure in a company and its divisions.

The information developments have been progressive	**As each stage of technology has opened up possibilities for new information processing arrangements, the potential has grown for improving the characteristics of management information.**			
Period	70's	early 80's	late 80's	90's
Processing				
Computing	Central	Central	Distributed	Common purpose central computer networked with special purpose distributed
Data Links	Physical	Electronic	Electronic	Optical
Data Storage	Access to fixed combinations only	Access to fixed combinations only	Flexible	Flexible
Information function				
Data enquiries	Cumbersome	OK sometimes	OK	OK
Flexible analysis	No	Cumbersome	OK sometimes	OK

PART TWO

Effects of the IT revolution on
business strategies

Groundchange

The IT revolution changes the importance of intermediaries and increases the economic advantage of integration between members of the supply chain

27

CORPORATE FRONTIERS

IT is threatening the middleman's past ownership of customer/supplier connections

Value added and data services (VADS) can help commercial and industrial customers and suppliers find each other. VADS allied to electronic ordering and funds transfer systems can help customers and suppliers open up and administer their relationships, and they can help them overcome the economic advantage that intermediaries may previously have had in doing this.

The capability to maintain and analyse very large databases makes it possible to build direct marketing systems which focus sharply on particular categories of consumers. Such direct marketing systems – now starting to be built on a very large scale – allow the supplier to bypass the intermediary.

Likewise, systems providing direct – and immediate – connection of consumers to suppliers are emerging. For example, Videotex mail order systems make it possible for the domestic consumer to shop through the technologies of video and telecommunications; their installation provides another route to what is being called "disintermediation".

Cutting-out middlemen through on-line information channels

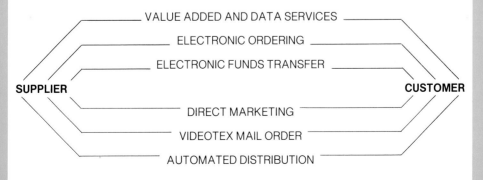

SUPPLIER — VALUE ADDED AND DATA SERVICES — ELECTRONIC ORDERING — ELECTRONIC FUNDS TRANSFER — DIRECT MARKETING — VIDEOTEX MAIL ORDER — AUTOMATED DISTRIBUTION — CUSTOMER

Electronic ordering systems bypass hospital stores	**Baxter – the US health products manufacturer – has established electronic communications links to carry orders direct from its customers' wards, clinics and theatres to its order assembly and despatch sites. Baxter now despatches direct to the users without having to go through the hospital stores; its share of hospital supplies has increased and funds locked up in hospital stocks have decreased.**
VADS threaten freight forwarders	**The freight forwarding business provides an example of how VADS help customers and suppliers, and how they threaten the position of intermediaries. Airline freight customers can now make direct arrangements for space on aircraft by using VADS to access airline in-house systems. VADS can also be used to simplify customs administration. Automated customs clearance and air-cargo forwarding give exporters more direct control over the service they can offer; but the position of freight forwarders is threatened.**
Direct marketing systems increase a car manufacturer's market control	**A major US car manufacturer has taken over a significant part of the management of customer connections by installing a database marketing system. This large computer system allows the manufacturer to keep in touch with people who have bought its products. By judicious use of the system to send messages to customers by direct mail, loyalty is built and sales are stimulated. In addition, the system monitors the service which customers receive from dealers. This helps to identify weak or disloyal dealers, so corrective action can be taken. The balance of power between manufacturer and dealer has been shifted towards the manufacturer, but both the manufacturer and the good dealers are better off as a result.**

IT changes the comparative economic advantage of the middleman's buffer role

The technologies of data capture (for example, through reading barcodes at point-of-sale terminals, or through smart cards), databases, and networks are making it possible for the details of customer requirements and deliveries to be captured and transmitted directly to the manufacturer or supplier for immediate analysis. The resulting information on the status of the supplier's stocks and on forecasts of demand can be more timely and more completely analysed than the information which can be inferred from transactions with intermediaries.

This direct availability of customer information can improve the economics of having the supplier maintain his own stocks and change his production schedules to cope with the economics and vagaries of customer demand patterns and locations. It can reduce the advantages of using an intermediary to hold a buffer stock to uncouple the supplier system from customers and to achieve distribution economies.

Will the middleman be cut out or will his power be increased?

In some circumstances, the technology can strengthen the position of the middleman. Which member of the supply chain has his economic power most enhanced by IT is linked to who manages point-of-sale customer activity and to whether competing products are available with equal convenience at the point-of-sale. For example, a big food retailer – an intermediary between suppliers and customers – can keep and use the information which it can now obtain from closely and quickly tracking the sales of competing products; it can thereby increase its power over its suppliers, and possibly absorb some of them. On the other hand, where circumstances are such that the supplier is able to establish a direct link with the customer, the advantage of the intermediary can be reduced or removed altogether.

But the advantages of using the new IT for direct contact between customer and supplier now provoke the question as to whether a middleman should be cut out from the supply chain.

A middleman can use IT to increase his market power

The UK insurance services subsidiary of the Automobile Association (AA) has installed a computer system which gathers data about a potential customer for car insurance – recording all the rating factors used by the various companies which belong to the AA's panel of insurance suppliers. The AA uses this information to automatically select the insurance company which is likely to provide the AA's customer with most appropriate and cost effective insurance.

This has strengthened AA control over the connections between insurance companies and their customers, giving the AA a tighter grip on the car insurance market as well as a major database through which to market other services.

Corporate frontiers are being redrawn to take account of disintermediation and integration

IT is changing the links in the chain joining customer to distributor to manufacturer to supplier.

Defining the best arrangement is a compromise. The compromise determines where the frontiers should lie between the different business units along a chain, and indeed how many separate business units there should be. Getting the compromise right is crucial for competitive strength. Total costs, market share, and adaptability are all at risk. The right compromise depends on the relative advantages of simplicity and complexity.

Prior to IT, simplicity won. Economics and practicalities led to the chain between supplier and customer being broken by intermediaries into pieces. The management of each piece of the chain was simpler, but at the cost of the suppliers losing some control over relationships with their ultimate customers and losing responsiveness to their customers' emerging needs.

After IT, complexity can be more readily coped with, and its relative advantages increase. One response is to cut out the middlemen, using IT to replace some of their functions, such as providing accessibility. A second response is to integrate members of the chain under one corporate management, using IT to help manage the detailed web of connections between supplier and customers, thus securing the potential competitive advantage of integration.

The benefits sought from disintermediation and from integration are many and varied. For example: the better contacts and responsiveness flowing from direct relationships; sharper end-to-end control and accountability for quality of the delivered product (particularly important for perishable goods); avoidance of confusion between objectives of the members of the supply chain; and lower costs. The price is the complexity and cost of the necessary computer and communications systems – but these are becoming progressively cheaper and easier to build and run. The price can also be the loss of the intermediary's window on competitor activities. And a question may have to be raised by regulating authorities as to whether monopoly is created by the concentration in ownership of information.

Coloroll takes over the customer/supplier connections and integrates

John Ashcroft, chairman of home furnishers Coloroll, knew what his company had to do to beat competitors. But his problem was that the traditional route of selling his company's wallpapers and fabrics through independent wholesalers and retailers did not allow him to do what had to be done at a competitive price. So, making major and creative use of IT, he broke away from the traditional route.

He wanted at the same time to show customers a wide range of coordinated designs, and to deliver within one or two days of an order. Using the traditional route, this would have meant high variation of demand on the manufacturing plant, very high stocks in the system as a whole, and very high costs, and even so it would have been difficult to ensure that the wholesalers and retailers made a reality of the brand image which Coloroll wanted to have for variety and service. Instead he established direct channels to some of the retailers, bypassing the wholesaler middlemen in these cases. Responding to telesales orders, goods were shipped direct from manufacturer to retailer. Level of service was kept high, costs were kept under control, image was maintained, and information was provided in sufficient quality to support aggressive and focused marketing and efficient production.

With its new strategy, Coloroll went from a £6 million family firm in 1978 to be the £35 million UK wall covering market leader in 1985. But that was not the end of the story. Coloroll started to reap economies of scale from the large investment it had made in its IT systems and its service to retailers. It took over other manufacturers, selling to their customers and extending its range to home furnishings, carpets, ceramics and glassware. It pumped more goods through the pipeline it had created, responding to the greater amount of information the systems were now collecting. Four years later, Coloroll achieved a turnover of £700 million, with Ashcroft saying "I want to create the largest home fashion group in the UK and then in the world".

Magnet has made a similar use of disintermediation, relying also on building IT into the customer service

Tom Duxbury, chairman of Magnet, makers and sellers of fitted kitchens and bedrooms, had a similar vision; to take direct control of the connection between customer and manufacturer and to make it work to the advantage of the customer and the company.

In the past, having a kitchen or bedroom designed and fitted could be a slow and frustrating affair for the customer. Understanding the customer's vision and matching it to reality were the first painful steps in the process. Attempts were made to use computers in some DIY stores to ease the process. The salesman

would develop a design and display it on a computer screen, and modify it immediately to fit the customer's ideas. But then came frustration following long delay in availability of some of the parts required for the design.

Magnet removed the frustration by taking the Coloroll route, cutting out intermediaries between manufacturing and sales and relying heavily on information to keep the process moving. Now the in-store computer screens – linked to information and the availability of parts – are a success. From being a lack-lustre company with declining profitability and margins and little sales growth, Magnet moved to sales of 75,000 kitchens a year, 60 percent through the computer, with an annual growth in profits of 20 percent, in readiness for a management buyout for about £600 million.

The capabilities of the new IT should stimulate supply chain members to review the frontiers of their activity

Management should re-evaluate:

- How are the practicalities and costs of engaging in complexity changing?

- What is the new balance of advantages between simple and complex arrangements?

and hence

- Should middlemen be cut out?

- Should there be backward or forward integration between members of the supply chain – and what takeovers or mergers does this imply?

in other words

- What business are we in? How much of the process should we control?

Intermediaries should produce a strategy for survival and power

In the face of the new opportunities for other members of the supply chain which are opened up by the new IT, management of an intermediary should review:

- Can technology be used to lock in customers by connecting them through information systems which will be too expensive or difficult for the customers to change?

- Can the quality of the information provided by an intermediary to members of the supply chain and the productivity of the intermediary be improved enough to justify retention of the intermediary role?

- What asset of the intermediary could be created within the existing structure which another member of the supply chain would want to take over?

- By exploiting his access to information could the intermediary initiate integration and be the owner of the resulting operation?

6

MARKET STRATEGY

IT shrinks distance and favours global operations

"What business should you be in?" leads quickly to the question "in which markets should you sell your products and services?"

The answer depends on how the barriers of geography and distance affect issues connected with time zones (for example, for 24 hour financial trading), distribution, product costs, and sensitivity to local matters. The markets which give you competitive advantage are influenced by whether geography favours you or your potential competitors, wherever they may be based. Much of the answer depends on how quickly and cheaply information can be transmitted and analysed. For example, a salesman wants to know how quickly and at what price an order could be delivered; whether from local stocks, distant stocks or straight from manufacture; and whether it is for a standard product or to meet particular requirements which might be shown in a drawing or picture.

In the past before IT and telecommunications, the answer would frequently lead to local markets. In the new world of global networks carrying sound, vision, text, and computer data, and of salesmen and schedulers having immediate access to common databases, the answer can change. Global networks now present an opportunity to businesses which are powerful enough to exploit global markets, global manufacturing, and global service. But to businesses without the means to exploit global networks, they represent a threat.

IT can be the means of raising barriers to market entry

A successful IT installation can create a barrier to entry for any competitor who is not prepared to invest what may be a considerable amount of money and effort in developing an equivalent IT system to improve performance.

IT may be used more dramatically to create barriers. IT can be used to add value to the service offered to a customer, who may then find that to take a similar service from a competitor will cost too much in duplicated hardware or systems or operator training. The customer is then effectively locked in.

Doing this successfully creates a barrier to entry. But timing can be crucial where the system depends on technical innovation. Trying to lock customers in too early carries the penalties of an immature technology. Doing it too late may mean yielding the market to a competitor. Getting the timing right can be key to market strategy. And vigilance will be required to spot when further IT developments would allow a competitor to get past the barrier.

Federal Express provides system backup to global operations and raises IT barriers to other distributors

Fred Smith, founder of Federal Express, has helped other businesses to lower the distance barriers to nationwide and global marketing. In so doing, he produced a growing billion dollar business in ten years without mergers or takeovers. Federal Express aims to differentiate itself from other parcel distributors by providing "time certain" deliveries. Critical to maintaining its "Absolutely positively overnight" reputation is its Customer Orientated Service Management Operating System (COSMOS). COSMOS enables Federal Express and its customers to keep track of each parcel in the distribution system. As well as using IT as an essential component of its market and product strategy, Federal Express raises barriers to competition for its high volume customers in the US by installing on-line computer terminals in their offices, joining them to COSMOS and capturing billing data.

IT can be used to overcome the complexity of variety

Coping simultaneously with activity in many markets involves complexity. Using computer models and support for the decision making process required to sort out the complexity offers a route to securing the benefits of market variety without the penalties. The difficulty of one centre coping economically with the variety of many markets can be a key factor in deciding the strategy for linking markets to a supply centre, and for amalgamating supply centres. Overcoming this difficulty will become important as barriers to trade within the European Community respond to "1992". As this happens, supply centres and markets will have to be replanned to take advantage of the detailed complexities of new possibilities for economies of scale.

Changes in IT may force reconsideration of market strategy

Management should therefore review the following broad issues:

- How could the IT revolution affect the connections of production centres with the company's markets?
- How are the resulting pressures for larger markets affecting institutional barriers to global activity?
- How are the effects of IT systems on economies of scale and specialisation affecting competitive position and product life cycles?
- What are the opportunities and threats for "locking up" markets?
- How do the maturity and power of the company's IT arrangements compare with that which the competition could have?
- In the context of the above, what should be the strategy for building and retaining the markets in which the company should compete?
  And this leads on to the overall questions
- What should be the associated strategies for products, manufacturing and distribution, human resources, and IT?
- What should be the associated strategy for corporate frontiers?

7

PRODUCT AND SERVICE STRATEGY

IT is starting to change the parameters of product variety in ways which are fundamental to the competitive strengths of product strategies

Getting the economics of variety right is key to a successful product or service strategy. The more a company can offer the right range of variety in its products or services, then the more responsive it can be to its customers, but at the price of production and distribution complexity and cost. IT is changing the economics of variety.

Large variety of product can be managed efficiently with IT based manufacturing and distribution systems

IT is reducing the cost of coping with manufacturing and distribution complexity. For example, IT supported administration and transaction systems give both production controllers and salesmen immediate access to the same stock and order information held in up-to-date databases, and give warehousemen the means to manage a large variety of stock efficiently.

And IT is helping companies adapt products more quickly and cheaply to meet customer requirements. For example, with the advances in distributed processing and very large databases, computer aided design (CAD) and computer aided manufacturing (CAM) can be used together, as CAD/CAM, to facilitate rapid product redesign and resetting of manufacturing. Furthermore, when the use of CAD/CAM is combined with automated assembly and warehousing to produce the so-called computer integrated manufacturing (CIM), the comparative advantages of product variety can again be increased.

The IT developments are all starting to make it possible to increase the amount of variety in product ranges which can be supported profitably so that greater variety can be included in a competitive product strategy.

Customisation to meet even the particular requirements of each member of a large market is becoming a practical way to increase market share – and indeed to increase the size of the market itself without the producer losing competitive advantage because of increased costs. Thus a mass customisation product strategy can now be an alternative to a global product strategy.

IT systems enable a diesel manufacturer to keep market share through offering product variety

One way that Cummins Engine Company, the major manufacturer of heavy duty diesel engines, has been able to maintain a position against heavy Japanese competition is by being able to respond quickly and economically to customer requests for many different varieties of engine. New management structures and automation in the factories were part of the means of achieving responsiveness. But also critical to the feasibility of its product strategy was Cummins' focus on using IT to provide on-line factory master schedules to support the production of large numbers of product variants and options.

Banks are using IT to become better at adapting their products

Banks are seeing that flexibility of product feature is a major competitive weapon in retail banking. A bank's ability to offer features such as flexible pricing, tiered rates, and flexible fund switching depends on whether its computer systems have been designed so that they can be quickly adapted to new requirements. To meet increased competition, coming not only from established players in the banking market but also, since deregulation, from building societies, UK banks have been increasing the focus of their product processing systems on flexibility of product introduction and enhancement.

Manufacturing firms are investing in CIM for many different reasons

An Alco plant in Tennessee is engaged on a $400 million programme primarily focused on making sheet metal of better quality for aluminium cans than its competitors can make.

In Italy, Fiat wanted to respond more quickly to customer orders. Being efficient and flexible was the key. Fiat changed to having twenty-five robots assemble twenty varieties of cylinder heads, with only 14 percent of the previous labour force. Computers track each cylinder head and engine block so that they are ready for engine assembly at the time they are needed.

Sandoz Pharmaceuticals, based in New Jersey, recognised that national contracts and buying groups were putting pressure on pharmaceutical companies for greater efficiency in manufacturing and distribution. Sandoz set out on a major CIM implementation to automate the very detailed material control tracking required for pharmaceuticals. The firm now runs its business with an integrated system, including materials requirement planning, bar coding, automated guided vehicles and an automated ware-house.

IT developments are tending to reduce product lives

The pace of IT development is increasing the rate of obsolescence of products which, like cars and cameras, contain information handling devices. This is one factor tending to reduce product lives.

A second factor is the high cost of establishing CIM. Because of its cost, a high enough volume of high value product is necessary to achieve the return on investment needed to justify CIM. The plant adaptability associated with CAD/CAM and CIM supports the maintenance in a competitive market of the high volume needed to justify CIM. As a result, there is pressure to keep up a flow of new and more competitive products to maintain overall volume of high value, and hence to reduce product lives.

These trends in product lives and whether the market place can absorb a high volume of high-value short-cycle products need to be taken into account in product strategies, with their consequences on product development and marketing.

IT supports investment banking innovations

Investment banking, with its intensive use of IT, has seen the rise of a new breed of technology-led financial product experts, whose sole remit is to use IT to develop rapidly new trading and dealing support products in the electronic markets. These so called "rocket scientists" have exploited to the full the hardware and software capabilities of the microcomputer to produce new sources of revenue on the trading floor very rapidly. The speed of product introduction is matched by the rate of product obsol-escence; the ability to compete depends crucially on the skilled use of IT.

IT developments are reducing the lead times for new products

CAD/CAM can have a major effect on reducing the lead time and variable costs associated with developing and launching a new product.

In manufacturing industries for which R and D is a long and expensive part of launching new products – such as aero-engines – replacing some of the physical testing of the performance and failure limits of critical components and materials by computer simulation is starting to have an important effect on the length and cost of development, and also on lowering the dividing barriers between design and manufacture.

In the service industries, the new products which can be offered to catch market opportunities are likely to depend on how quickly and cheaply the administrative backup computer systems can be designed and implemented. New system development methodologies and tools are available to do this. They make up the new computer aided software engineering (CASE); this is based on the powerful computer workbenches that are now available to systems developers. As well as improving quality, CASE can very significantly improve systems development productivity and lead times. Recognising this capability can be crucial to developing competitive strategies for services in time to catch a window of opportunity.

Volvo and Jaguar Cars reduce the lead time for new products through computerisation

Volvo has been working hard to shorten the design and development process. Its Dutch operation has been streamlining the process. A key contributor to improvements is a computerised data management system; this ensures that designers and engineers do not have to waste time finding technical data. Another key contributor is CAD; this cuts into that part of designer time – about a third – which has traditionally been spent on design. Volvo is looking to reduce its development and introduction lead time for a new car by 30 percent, and to make even bigger percentage reductions in the time for upgrades.

Using a similar approach, Jaguar Cars reckon to have reduced the lead time for taking a luxury car from concept to production by about 20 percent.

IT makes it possible to choose more effective product strategies

A product strategy involves complexities of timing, finances, marketing, sales and manufacturing. Getting the details of these complexities right can make the difference between a product strategy which succeeds or fails. Working out the details for various alternative strategies can be such an onerous task as to limit the variety of alternatives explored. This difficulty can be overcome through using one of the available "user friendly" computer systems which are designed to simulate the consequences of different strategies. This computer modelling can make it easier to choose an effective product strategy.

The choice of the most profitable products to include in a particular product strategy can depend crucially on fast and accurate assessment of direct product profitability. This requires bringing together current data from many different sources; it is clearly a task for on-line computer systems.

Financial institutions model alternative product strategies	**Financial institutions now face markets which have become much more competitive since deregulation. They have used direct product profitability modelling to review potential products and, in particular, to assess the delivery channels for products to ensure that the most effective delivery strategies are selected.**
Retailers adjust product strategies using up-to-date computer data	**The Boots retailing chain gets current data on its sales of 60,000 items using barcode scanning equipment in many of its 1000 stores. This data is combined with data from buyers, store management, warehousing and distribution to give direct product profitability data; the combined data is used to manage the selection of items to stock and promote. Increased profit in the stores using the system is said to have resulted.**

Total customer service can be provided by using IT to integrate different parts of the service chain

To provide best possible response overall to customer requirements, the various parts of a service providing system need to be tuned to work together. Such an integrated service can sometimes be seen as a new service, and one on which, in effect, a new business can be built. On-line processing, with access to large and remote databases, is the prerequisite to such service.

Systems to support integration are a key component of NHS strategy for total consumer service

A strategy has been devised for the UK's National Health Service (NHS) to provide people with a computer supported integrated service from general medical practitioners (GP's) and pharmacies. If all elements of the strategy are eventually adopted, several stages of service to the patient will be integrated, leading to improved patient care and lower NHS costs. A patient's "smart card" would be coded by the GP with prescription details; the card would then be encoded by the pharmacist with dispensing details which would simultaneously be transmitted to a central computer for NHS pricing, achieving some cost savings on the way. The card would build up the patient's medication history. The availability of this history in computer readable form would help the GP and pharmacist to check potential drug interactions, and help the patient's care in emergencies.

Another change envisaged is to link a central register system electronically with the databases of patient GP registrations held on the computers of Family Practitioner Committees. This will make it much easier for a patient to change the GP with whom he or she is registered, the time taken to transfer medical records from the one GP to another going down from three months or so to one or two weeks. This should help strengthen the ultimate power of the consumer in the new NHS arrangements envisaged in the 1989 government review of the NHS in which the GP would act as an agent for bringing NHS facilities to the patient.

The competitive advantage of an information based service may depend on exploiting the possibilities of IT

IT developments change the feasibility of giving customers access economically to information based on vast amounts of up-to-the-minute data. This changes the expectations of customers for the information and its benefits. IT now provides, in effect, the machine tools of the information industries such as banking or stock markets, and it is critical to their strategies.

Management should produce a product and service line strategy in the light of IT developments

In considering their product and service lines, management should review the following broad questions to assess how their positions and how the positions of their global competitors are being changed by IT developments.

- To what extent would a faster or more varied response to market opportunities give competitive advantage?

- How would better administration and transaction systems improve the ability to respond?

- To what extent would CAD or CAM or CIM or CASE provide what is needed?

- What would be the costs and lead times of implementing these IT developments? When would it be realistic to base a product strategy on them?

- As competitors adopt these developments, what effect will there be on product life cycles and on the resources required for developing new models and products?

- Are sufficiently complete evaluations of alternative product strategies being made?

8

Putting IT into the production/distribution chain can support
increased sales and profitability by making it easier to improve
product range and availability and to lower operating costs

47

PRODUCTION AND DISTRIBUTION STRATEGY

Strategies for production and distribution are developing in two different directions depending on which of two approaches to managing information is more appropriate

One approach seeks a return from investing in the capture of detailed data on product movement and using the data to manage stocks held at various parts of the production/distribution chain.

By contrast, the other approach seeks to avoid stocks in the chain. It relies on planning the chain and scheduling its parts so that there is efficient response to variations in product movements.

Production and distribution strategies may need to be changed to exploit the product tracking capability provided by new data capture methods

New electronic methods can capture detailed information on product movements as a by-product of normal operations (for example, through barcoding on products being read electronically at check-out tills). This opens up possibilities of detailed tracking of products through the production/ distribution pipeline and of computer based scheduling methods to take best advantage of capacities and demands. This allows the system to maintain better stock availability with lower stock levels – and hence produce larger sales and better margins. The availability of the better data which is produced supports the "disintermediation", or cutting out the middlemen, which was discussed in Chapter 5 and supports the "Computer Integrated Manufacturing", which was discussed in Chapter 7.

Toyota uses systems to track cars through production and distribution to achieve very rapid delivery of orders

Toyota found that more costs and more time were incurred by a car in the distribution system than in the factory. The manufacturing company absorbed the sales company, linked the manufacturing and distribution control systems; and Toyota today is working to deliver cars to customers within a day of receiving a dealer's order.

A computer manufacturer bases its spare parts distribution strategy on an automatic fault communication system

The spare parts distribution strategy of Stratus Corporation (computer manufacturers) is such that when a customer's computer has a fault and the backup takes over, the factory automatically picks up a signal from the computer. The first time the customer knows anything is wrong is when the replacement part arrives next morning.

A just-in-time strategy for production and distribution avoids investment in information on stock movements but may require investment in factory information and scheduling

In the approach of "Just-in-time" (JIT), production and distribution events are initiated by physical flows of items rather than by data about the physical flows. JIT aims to link different stages in the product chain without the buffering between stages which has traditionally been provided by stocks. The change in arrangements is fundamental. Manufacturing lines may have to be reorganised; factories may have to be regrouped and resited; staff may have to be retrained. The practicality of JIT is affected by the ease with which the producers in the pipeline can respond to variable demand; this capability may depend on computer support of plant adaptability and of precise scheduling of key units of production capacity.

For survival in a world where competitors have changed their physical and IT arrangements to achieve the benefits of JIT, management are obliged to consider whether and when JIT should be part of strategy. If they do decide to adopt it, then they have to recognise they are in for a period of fundamental replanning of their arrangements. This can be assisted by using the computer to explore alternatives. And they are also in for a period of redesign of their computer systems to concentrate on plant scheduling rather than on the materials requirement planning of the pre-JIT world.

Whether or not to introduce JIT is clearly a strategic question. But, when JIT works, stock costs are saved and customer service improves. The benefits of the change can be much greater than the costs.

Management should reconsider their strategies in the light of new production and distribution philosophies and of IT possibilities

Management should review answers to these questions:

- Should detailed data about product movements or the product movements themselves be the triggers to action?

- How would investment in securing detailed data affect broader issues of corporate frontiers and manufacturing investment?

- What should be the role of IT in any fundamental realignment of activity according to JIT principles?

A 'Human Resources Strategy' must take account of IT because the frontiers between people jobs and computer jobs are changing

51

9

HUMAN RESOURCES STRATEGY

IT changes people's jobs of thinking, communicating and doing

Changes in people's jobs are implicit in the changes in organisation and strategy induced by IT and highlighted in this book. This chapter highlights some of the people issues which may result from changes in jobs and for which management will have to think out a strategy.

Recruiting, location, training, career paths, rewards and management style are all affected directly or indirectly by IT. So also is the balance of power, both to manage and to disrupt. The people issues impacted by IT are broad and deep – and critical to success; they can only be touched on by this chapter's agenda of IT implications on the way people carry out their jobs of thinking, communicating, and doing.

Small numbers of exceptional problem thinkers will replace large numbers of standard problem thinkers

For sustained competitive strength in a world impacted by IT, an organisation's human resources strategy must cultivate a shift from large numbers of routine "standard problem" thinkers to a core of "exceptional problem" thinkers.

Consider the activities of thinking. The activities are: gathering, observing, and analysing facts; creating and testing ideas; and deciding what to communicate and do.

By exploiting the power of computer networks and databases, the thinker can gather more and better quality facts, both from within the company and from external sources, and he can do this in less time, with less effort, and with less difficulty. Similarly, by exploiting the presentation capabilities of micro-computers, the thinker can observe the gathered facts more easily, more selectively and more graphically.

So, provided the analyst is capable of exploiting the extra power available to him, his **analysis** of the facts can be more powerful. But some of the analysis is likely to be routine and "programmable", and need no longer be done routinely by humans. Then there is less room for the "standard problem" analyst, but there is relatively greater need for the "exceptional problem" analyst able to do the difficult parts of the work left by the computer.

A similar argument applies to creating and testing ideas and deciding what to do. This suggests that, to do its work as well as is possible in the IT age, an organisation needs fewer – but better – thinkers. The argument applies one way or another at all levels.

The activity and resources of recruiting, the rewards of staff, and the training and style of management have to be assessed in the light of this.

A human resources strategy has to take account of how the IT/communications revolution is changing the distance barrier to communication

A human resources strategy has to take account of the new flexibility in where people should be located to do their work. It may also have to take account of an increased global reach for the work.

The strategy has to exploit the potential advantages of people working from home, or in a network of small and dispersed offices, linked electronically for voice and data and – through means such as videoconferencing – for visual human contact as well. The strategy has also to recognise and cope with the human problems such arrangements can create. Among these problems are the effects on supervision and leadership which can follow from the remoteness and loss of personal contact resulting from replacing face to face contact by face to screen contact. A human resource strategy needs to recognise these new problems and find ways to lessen their effects through creating compensating occasions for human contact.

IT has a major impact on the arrangements for changing the way people do things in their jobs

People do things, physically, in the time remaining after they have finished thinking and communicating. For example, they physically document the results of their thought and communication, or they move things, or set machines.

The impact of IT is on the quantity of such tasks left for them to do, and on the signals they receive to guide them in what to do. The human resources strategy has to recognise the training required to cope with the amount of change in working procedures that can result from all the changes stimulated by IT change. The speed of change that may be required and its size can introduce a new level of challenge to training facilities, and this has to be recognised.

The response of UK clearing banks to IT requires major reorientation of staff

The large UK clearing banks were at the forefront of automation in the 1960s. Computer systems were developed to automate much of the manual accounting carried out in the banks' branches. This automation was successful; together with various cooperative efforts (for example, centralised cheque clearing) automation enabled the banks to increase the volume of their activity without increasing staff at the same rate. However, the organisation structure of the clerical work forces was largely unchanged, with a strong hierarchical structure repeated through many thousands of branches.

Deregulation is now driving change. The increased penetration of building societies and retailers into the financial services industry generally and into banking in particular has made the marketplace increasingly competitive. The clearing banks are facing an enormous challenge which they must meet by improving profits through reduced costs and increased revenues.

This requires a fundamental cultural change, to move from the traditional focus on administration and control to a sales orientation. IT will be at the heart of this change.

There will be a fundamental redesign of the banks' basic systems to automate many of the evaluation and control processes and streamline further the administration procedures. Many of the traditional roles of branch management will then be automated or de-skilled. Also, systems will become customer orientated rather than account orientated. The results of these changes should be that staff will be freer to concentrate on serving customers and systems will help them to do this well. Major reorientation of human resource strategies will be needed.

Sustaining morale and motivation of staff is critical to achieving the benefits of IT

Installing the products of the IT revolution may greatly change people's jobs and work relationships. Depending on how it is managed, the process of change may help or hinder morale and motivation of staff, and hence how well they think, communicate and do. Getting the change process right is critical to ensuring that the intended benefits of change are achieved. Getting it wrong may not only mean that the projected benefits of change are not achieved, but, worse, the vulnerability of an organisation to its computer systems may be exploited by a disaffected employee tampering with a system's operations or information.

Planning the processes for managing IT induced change in jobs is key to a human resources strategy. This subject comes up again in Chapter 12.

The Department of Social Security uses IT to train staff to operate its new computerised social security system

The Department of Social Security is making major changes to the way it uses computers to calculate and pay out one billion pounds of benefit entitlements in Great Britain every week. As a result, 45,000 staff of the Department are changing what they do in their jobs. In future, their jobs will be centred on their clients and their computer work stations. To change their work methods requires a massive training exercise. The training is itself based on computers. Trainees use a computer which explains new activities, takes them through the activities, and then provides practice and checks that they have mastered the systems and new ways of working. The use of computer-based training should ensure that the training is carried out consistently, with less disruption, and in time for the new system to be operated from offices across the country.

Management ought to consider the impact of IT change on their strategy for human resources

In the light of the above observations, management should address broad issues such as:

- What change will be needed in the number, location, calibre, and skills of people?

- How will the work environment change, and what needs to be done as a result?

- What changes will be required in people management and in support activities?

- How can the security of the organisation be protected?

PART THREE

Effects of the IT revolution on
organisation structure and
management

Teamchange

10

MIDDLE MANAGEMENT

Middle management organisation needs review

Top management has always had a concern to see that middle management is properly organised. Without good arrangements of middle management's roles, responsibilities, relationships and people, the implementation of top management's strategies becomes frustrated. And without good middle management arrangements, adaptation of the strategies to change in markets and technology is slower than the competition's.

Middle management effectiveness and cost – often up to 15% of product cost – are key to securing a strategy for competitive advantage.

The continuing revolution in IT and in the possibilities and costs of processing information have opened up new and important challenges for improving the productivity of middle management. This chapter aims to help top managers focus on key questions about middle management's future role and arrangements in the light of the technology revolution.

The developments in IT which are key to the arrangements for securing middle management productivity are in distributed computing, communications networks, database management, systems development methodologies and user interfaces, all of which have been described in Part One.

A number of key issues and jobs are worth reviewing

The sources of corporate competitive advantage which are affected and may offer opportunity for improvement are:

- Personnel cost.

- Adaptability to changes in policies and in the corporate environment.

- The tightness of control of efficiency and effectiveness.

- The quality of problem solving.

- The quality of leadership of the workforce.

  Questions relating to middle management are:

- How many levels of management are appropriate?

- What separate functional units are appropriate; where should department frontiers be drawn?

- How many middle managers are required?

- What competence and calibre are needed to do the changed mix of work?

To answer these questions in the context of IT developments, top management should review the jobs in which middle managers and staff:

- transmit and filter top level directives to lower levels of activity and monitor their productivity; that is, middle management's *relay role*.

- take decisions for lower level activities; that is, their *problem solving role*.

- coordinate activities and planning with other departments; that is, their *functional and coordination role*.

Middle Management's relay role is changing to require relatively fewer people

The relay role requires middle management to interpret and focus management directives for lower levels of activity, and then to review performance, comparing it with standards and between different centres of the activity.

Traditionally, this is done by personal contact, perhaps with the aid of some reports of costs and output. Management time is taken up in collating information and in "walking around". The smaller the time required, then the larger is the practical supervision ratio.

As the technology of distributed processing and communications facilitates the recording of information on activity at or close to the point where the activity takes place, and makes the information readily available for analysis by managers located away from the activity point, so less management time is needed for the relay role, and supervision ratios can be larger.

Furthermore, performance has the possibility of being made more effective, because the available performance information has better timeliness, accessibility and computability. Management can be tighter, with effective delegation being achieved in a tight framework of performance feedback. Delegation, without abdication of responsibility, becomes easier – even with relatively large supervision ratios.

So not only does IT support for the relay role lead to fewer people but it also opens the way to reducing the number of management levels. The benefits of such simplification in management structure to overall productivity and adaptability can be large. Indeed, it has been observed that large US companies with good long-term financial performance have, on average, fewer headquarters staff and almost four fewer layers of management than those with poor performance.

Peter Drucker: Harvard Business Review (January 1988)

"Whole layers of management neither make decisions nor lead. Instead their main, if not only function, is to serve as relays – human boosters for the faint, unfocused signals that pass for communication in the traditional pre-information organisation."

Middle Management's problem solving role will become relatively more important

The new technology is changing the frontiers between levels of activity.

The nature of the lowest levels of activity is changing. With computer support, the shop floor worker can now be responsible for a wider range of manufacturing tasks on a larger volume of production. Likewise the salesman can be responsible for more of the job of making a delivery and price commitment to a customer. As a result, the training and calibre of people at the lowest level doing these enriched activities needs to be higher.

It is then to be expected that middle management will receive a new mix of problems which cannot be answered either by people at the lowest level or by computer. There will be relatively fewer easy problems and relatively more difficult problems for middle management to solve.

Just what is the effect on the time and the qualities required to do middle management's job depends on circumstances.

Contrary to the relay role's effects of increasing supervision ratios, the effect of the problem solving role can be to decrease ratios, but over a smaller number of lowest level people. In other words, the effect can be more output per person at the lowest level, fewer people at the lowest level per supervisor, and fewer supervisors per unit output.

Likewise the calibre of each middle manager – now responsible for much more output – may have to increase. And, possibly, the competence and knowledge which the middle manager requires may be indistinguishable from what is needed at the level just above him, with the implication that the two levels should not be distinct. So the number of management levels can again decrease.

Furthermore, to increase the effective problem solving capability of middle managers, to release their collective energy to help each other to solve problems, the new technology opens up the possibility of building networks to facilitate problem and data sharing. Using such networks, no longer will a middle manager have to use his energy to pass his unsolved problems up the organisation before the problems come down again to other parts of the organisation which may be able to help.

Elopak finds its problems faster

Before you can treat a problem as an opportunity, you have to know there is a problem. Euilf Storm, chief executive of Norway based packaging company Elopak, knew that he needed to know more and sooner about customer and product problems if he was to exploit opportunities for improving customer service. He knew he had to do this if he was to cope with powerful global competition.

Elopak has a salesforce and a technical maintenance force to make contact with customers. Its salesmen meet customers from time to time to sell Elopak's huge packaging machinery and the packaging materials for the machines to use. But its technical people call regularly to maintain the machinery. Euilf Storm saw that if the salesforce were to be able to nip a customer problem in the bud, then they needed to know quickly if the customer started to buy more packaging materials from competitors. The salesforce and machine designers also needed to move fast if machines in service started to run into problems. He reacted by setting up computerised information systems.

One system kept track of packaging materials sales for each customer, comparing them to what was expected for that customer. The technical people were instructed to gather sales related information on their regular visits to a customer so that the expected sales profiles for the customer could be updated for any change in circumstances, and so that the monitoring of sales could be against an up-to-date profile. In this way, without adding to their size, the salesforce could be more involved in post-sales operations experience and could be more closely focused on customer relations problems as they appeared.

Another system kept track of machine problems for early warning to machine designers and to maintenance planners. More of the time of the salesforce was then spent on solving customer problems rather than searching for them. Also the technical force had to be upgraded so that it could communicate effectively with the higher levels of customer management who could give them the sales and competitive information they needed.

The management horizons of both salesforce and technical force were broadened, and the calibre of people in each had to be higher. The result has been that Elopak has been able to maintain its progress against strong global competition.

Reconsideration of middle management's relay roles and problem solving roles should prompt some basic organisation questions

In the context of new IT possibilities, top management should review these broad questions to assess their implications for increasing productivity in the circumstances of their organisation:

- How does computer support of the lowest levels of activity and of the relay role between top and bottom change the number and skills of middle management?

- Do supervision ratios therefore need to go up or down? What happens to spans of control?

- What scope is there for reducing the number of management levels because of IT support of the relay role, and because increased problem solving capability at the lowest level reduces "head room" between levels?

- What scope is there for reducing the number of people needed to assemble relay data?

- What scope is there for improving corporate adaptability (and for keeping top management in touch with the reality of problems) through having fewer filtering levels of management and through having more timely information about "sharp end" activity?

- What scope is there for improving morale through better information helping management to avoid unproductive dialogue between top and bottom about what actually happened?

- What activities to facilitate human relationships should be established to compensate for reduction of face to face contacts?

- How should the selection, development, training, incentives, and career paths of middle management change?

- What changes are required to manage the new style middle managers effectively?

- What computer and telecommunications support should be provided for middle management's new responsibilities?

- How can the implementation of organisation changes be timed to keep in step with the implementation of the technical changes?

IT developments are making it possible to change or remove the frontiers between operational departments and functional and support departments

What departmental structure should there be for the people responsible for operational activity such as manufacturing or sales, and for functional matters, such as finance or credit control or stock control? How should these people, with different skills and orientations, be split into departments? Indeed, to what extent are specialists still required for functional matters, and are functional departments required? An organisation's answers to these questions depend on the nature, complexity, and size of its business and, crucially, on the maturity and power of its information arrangements. The answers define the number of departments and their size and scope. In turn, these define the number and responsibilities of middle managers. As IT changes information arrangements, so these answers are changing.

Traditionally, people with functional skills tended to be placed in functional departments separate from operational departments

In the past, each department tended to carry out its own data capture, data processing, and information reporting. Because of the limitations on the capability, flexibility and economics of IT, and because it was the way things had usually been done, each functional department and each operational department had generally to have its own information arrangements. Each department used its own expertise both to assemble and to interpret its own information. People with functional skills were responsible for both the activities of solving functional problems and of processing functional information; and the two types of activities were placed together in a department corresponding to each function. Likewise for operations. And each department spent much time and effort in communicating, coordinating and reconciling its information and actions with those of the other departments.

Furthermore, economics generally dictated that a functional department served several operational departments, thus adding problems of separation between central and field departments to problems of separation between functions and operations.

Now IT developments enable systems for functional information and operational information to be integrated

No longer does each department have to manage its own information. As distributed database management systems and tools for network management and software development become practical and economic, the integration of data capture, storage and usage becomes practical. Data need now be entered into an integrated system only once, at the point where activity takes place, and then made available to its different users in different departments. The costs and confusions of data duplication are avoided. However, there is a price to pay in terms of the effort required to maintain accuracy of the data and its credibility with all its users.

As the economics and practicality of developing complex software improve, expertise of different functions can be built into one integrated set of computer programs. This enables the computer to make routine interpretations of data relating to different functions, combine them, and use them to make routine decisions. In addition the computer can present human decision makers with the information they require for assessing exceptional problems.

The more this computerisation of the routine can be done, and the more that functional expertise can be captured in so-called expert systems, then the less is the remaining day to day work for functional experts. Their contribution to existing routine work is progressively eliminated as systems incorporating their knowledge are designed and built. What remains for them is the non-routine work involved in functional planning and in exceptional problems.

Some financial institutions are exploiting information technology so that they can focus on products and they are redistributing functional activities accordingly

The ability to integrate functional and operational information has raised the prospect of greater organisational efficiencies in the retail financial services industry. Two trends are forcing organisational change: rapid diversification of products and services and increasing pressure on profits. Some financial institutions are therefore seeking to reorganise along product lines, moving away from traditional function-led structures (such as marketing or finance).

With the development of much improved sales, marketing and financial databases (both centralised and distributed) and investments now being seen in office communications and electronic delivery systems, there is the opportunity to scale down large central functional departments, and to transfer such functional support out to product management. Major financial institutions such as Barclays Bank and the TSB Group are pursuing this strategy. They are seeing their way to achieving leaner and much more efficient control and support structures. Through being able electronically to communicate with and control product-based initiatives, these structures will also be more effective. Product managers will become more effective as they extend the scope of their responsibilities, having direct access to functional information and using that information with the assistance of the de-skilled functional expertise built into the systems. The product managers will need to have the calibre to become business managers.

IT developments also allow systems for clerical support to be removed from the responsibility of function and operations managers

In the interests of overall job quality, it may be desirable to keep the responsibility for high volume routine clerical work (for tasks such as order entry and any data compilation which has not yet been computerised) separate from the responsibility for identifying and solving functional and operational problems. The skills and management styles required for the two types of work are very different.

The integration capabilities of modern IT and network management tools now make it possible to separate the management of integrated operational and functional work from management of the routine work which still requires human intervention. Managers no longer have to try and operate in two different modes simultaneously. Furthermore, it becomes possible to obtain economies of scale from central processing of such routine work as can be standardised.

There can then be a large effect on opportunity for those middle managers concerned with back-office support.

Middle managers are winners and losers in the game of Department Musical Chairs opened up by IT

IT developments make it possible to change company departments in new and productive ways. Appropriate activities of functional departments can be put into separate clerical units. And, as it becomes possible to computerise and integrate functional and operational information systems, other functional activities can be linked more closely with activities of operational departments, either through functional activities being taken over by operational departments, or through establishing better information links between departments.

Playing the game of Department Musical Chairs, illustrated in the diagram, department managers may expect to move out of the Traditional Stage – with separate operational and functional departments, each concerned with its own problem solving and clerical processing – into manoeuvring for the fewer chairs of the Transitional Stage, before struggling for the last chairs of the Ideal Stage.

The Transitional Stage of Department Musical Chairs exploits IT to increase the potential for corporate performance

In the first round of the Transitional Stage of the game, IT enables the problem solving and clerical information processing activities of a department to be split advantageously into separate departments. As well as opening the way to economies, this change is aimed at sharpening management focus, both in the resulting clerical unit which takes over the separate processing of what may be several departments, and in the remaining work of the departments. Pressure for the change comes from the belief that a department is likely to have higher performance potential if it does not have to encompass the different skills and management styles required for both high volume routine work and the application of specialist expertise to solving problems.

Then comes the next round of the Transitional Stage. Freed from its clerical responsibilities for information processing, a now slimmed down functional

Department Musical Chairs

Stage	Operational departments	Functional departments

TRADITIONAL

TRANSITIONAL

IDEAL

department becomes a candidate for its activities to be taken over by operational departments.

What determines the choice of whether functional activity is taken over by the operational department or the functional department retains its identity? Cost savings and the availability of managers suitable to run combined departments are important. But critical to the choice should be an assessment of what is involved in producing the appropriate supporting systems. To absorb activities of the functional department, it must be possible to build enough functional expertise into integrated information systems to provide functional support to operational management decisions, and to build a system of performance indicators to help control the overall quality of those decisions. Otherwise it may be better to keep a separate functional department but to link it closely with operational departments through effective systems.

The Ideal Stage of Department Musical Chairs depends on a high degree of integration and IT systems support

As the pressure from information developments builds up and systems become more powerful, so the game moves into its Ideal Stage. Then the number of established separate departments may be expected to decline further. In addition, the feasibility and richness of information systems available to a department in this stage should support adaptation to new challenges by making it possible to set up temporary task groups of people with functional and operational skills from within a department, focused on a particular issue, and accountable to the manager of the department. Also in this stage, the separate activities taken over by the clerical unit may be expected to be integrated and automated, with consequential reduction in numbers of personnel.

Better focus, adaptability, accountability and economy are the aim of changing operational and functional department frontiers

An objective of exploiting IT and indulging in Department Musical Chairs is to improve integration of functional and operational activities so as to sharpen and consolidate the *focus* of jobs on customers or products, without letting reorganisation cause unacceptable weakening of functional skills.

A second objective is to improve *adaptability* to external change and the unavoidable uncertainties of operations. This adaptability is key to sustaining any competitive advantage that a company has. But if separate departments are responsible for identifying and solving problems with a mixed functional and operational content, such as, for example, problems about linking price discounts and payment terms, then adaptability is likely to require difficult co-ordination and to be slow. Therefore, by using IT to make functional expertise more accessible to operational departments, bringing departments together and close to external activity, adaptability can be improved.

The third objective addresses *accountability*. Where accountability for profit or service performance is divided between separate functional and operational departments, then – except where division is necessary for internal control – it is likely that ability to compete or achieve value for money is thereby weakened. Therefore the third objective is for operational departments to absorb functional activities so that accountability for the various aspects of operations should be undivided.

The fourth objective is to improve *economy* by reducing duplication and the number of departmental managers.

A utility is taking advantage of IT to focus on customers and to lessen the job of coordinating different departments

A large UK utility – responsible for services to some 10 million people – is sharpening its focus on customers. In the past, a customer needing to contact the utility was likely to have been confused as to which of many separate offices he should contact. And the offices themselves were confused by which of many separate computer systems held information relevant to the customer, and how the separate activities of the offices interacted with the customer. As well as diverting management focus from customer service and confusing management accountability, the past arrangements were not well suited to preparing the utility to adapt to new commercial and regulatory pressures attendant on privatisation.

The response has been to work towards a view of a future intended to realise the service and profit potential of management focus, accountability and adaptability. Customers will contact only one office, whose staff will be trained to focus on customer service and whose systems will provide integrated information on revenue and operations. IT will be used to connect the integrated customer office with operating departments. No longer will a customer have to worry about who should deal with his query. No longer should middle management have to spend large amounts of effort and time in attempting to coordinate the activities of the different departments and in trying to ensure that the utility is consistent in its customer treatment.

The extent to which separate functional departments should remain depends on whether IT developments make it worthwhile to improve *corporate system performance potential*

Making it possible to change the corporate system – that is, the organisation structure and information systems – so as to improve the focus, adaptability, accountability and economy of departments, IT has the capability of increasing and sustaining the potential of the corporate system for generating profit or service. In other words, IT can be used to improve *corporate system performance potential.* This improvement in corporate system performance potential is a major objective of using IT to change middle management's functional and coordination job.

When is it economic to exploit the possibilities of IT and disrupt existing functional arrangements to improve corporate system performance potential?

Answering this question requires careful analysis and good judgement about the value of future possibilities and the costs of change.

An assessment of the likely overall value of changing the corporate system performance potential should be compared with an assessment of investment in the costs and management burden involved in changing the corporate system, and, in particular, the arrangements for information processing. The question that has to be resolved is whether the additional size and profitability of the markets that can be made available because of the increased corporate system performance potential, and the improvements in economy that can be secured, are such that there will, with management commitment, be a large enough return on the effort and expense of changing the information arrangements. This is a key judgement for top management to make with regard to IT, and only they are in a position to make it.

Girobank is taking advantage of new systems to improve performance potential

Customer services at Girobank in the UK used to be provided by many specialist departments. Because of the frequent need for co-ordination between the departments, it was difficult to provide a highly responsive and competitive service.

Taking advantage of the fact that their computer systems were due for replacement, Girobank management decided to exploit modern technology and build replacement systems which would improve customer service and workforce productivity.

The new systems guide clerks through various specialist enquiry handling routines and help them deal with many of the tasks previously the province of some of the specialist departments. Clerks now have accountability for adapting the action on a particular customer to his circumstances. Also, decisions which exceed the limits of their authority are now passed automatically to supervisors.

Clerks can now handle many decisions previously handled by centralised functions. Customers are allocated to account teams, enabling improved customer focus. Clerical jobs have become more interesting; supervisors are able to focus their activity more sharply.

Better customer service has resulted from the change in the frontiers between functional and operational departments. Jobs have been enriched. The new adaptability, accountability, and management focus have improved corporate system performance potential. The price has been a marginal increase in the cost of information processing above what a simple replacement of the old systems would have incurred. But the number of customers each member of staff can support has increased.

There will be less need for functional departments to exist for day to day activities and more need for line managers to have a broad business training to deal with functional problems

The trends in the critical factors all point to the at least partial demise of functional departments, and lessening of opportunity for functional middle managers. The trends are downward in the costs, upward in the force of competition, and upward in the amount of decision supporting functional expertise which can be captured in computer programs.

The change may be expected to happen first for departments with relatively mature systems – such as finance departments. It remains to be seen whether the change will happen for departments – such as logistics departments – which have come into being to coordinate operational departments, exploiting the new capablility of IT to provide information on activities in departments, but whose maturity has not yet progressed to the point where systems are integrated and able to carry out much of the coordination activity.

The manager of the new style operational department will need a broad business training and education to be able to combine sufficient functional expertise with his other skills to manage multi-disciplinary teams or cope with the exceptional "non programmable" problem without requiring the assistance of a functional expert in his department.

New systems enable Shell UK Oil to decentralise some functions to business units and to facilitate coordination between units

A central accounting service in Shell UK Oil used to price and invoice the company's sales. Taking advantage of new computer arrangements and responding to competitive pressures, the company has now decentralised the pricing and invoicing functions to individual business units; the result is that pricing and invoicing routines can now be more readily adapted to the needs of each customer.

Also, exploiting new database capabilities, Shell UK Oil has strengthened the role of a central coordination department in order to achieve better management of the organisationally separate activities – crude supply, refining and distribution.

Management ought now to question the competitive strength of their arrangements for functional activities

Top management should now ask how far the changed possibilities for functional work apply to their own operations, and what scope there is for the prizes of

- Fewer "turf" battles between departments.

- Sharper management and job focus.

- Sharper accountability, as responsibility ceases to be shared with separate functional departments, and greater adaptability.

- Less duplication of data capture and processing, and less work taken up in clearing up confusions of different versions of the same facts.

- Fewer (but better trained) managers.

Management ought now to review middle management's functional and coordination roles in the light of IT developments and get answers to broad questions such as the following:

- How do the economics of increasing the "corporate system performance potential" weigh up against the economics of changing the corporate system?

- Is it feasible to absorb previously distinct functions into line departments?

- How should any redevelopment of systems be phased in with restructuring?

- How much more sharply could accountability be focused and rewarded?

- How should staff be developed so that they can work in multi functional teams within the new departmental frontiers?

- To what extent are middle management functional jobs needed to build people for top management functional responsibilities?

- Are the answers to these questions being regularly reassessed in the light of business and technical developments?

11

TOP MANAGEMENT

The IT revolution is changing top management's job of formulating business and organisational strategy

From the chapters on Groundchange, it should be clear that the effects of IT on corporate activity can be profound. Indeed the effects may make it both possible and competitively necessary to push corporate activity into quite new areas of business and organisation. Top management therefore needs vision to see these new areas which can be opened up by IT.

Creating this vision is a task which requires all top management's experience and imagination. Fortunately the task is not made even more difficult by having to be confident about the details of what can be done with IT. This is because the trends in IT are such that there is progressively less need for the technicalities of information processing to interfere with the process of creating a vision of a practical future.

It now starts to make sense for top management to engage in what may be called *open information visioning.* The purpose of such an exercise is to envisage a future which is liberated from practical constraints on the availability of information about past and present activity and which accepts radical departures from current activity if they can be seen to improve factors of the business which are critical to its success.

In this way, without detailed knowledge of IT, top management can open up the process of formulating strategy to set long term goals which should exploit the trends of IT developments. But their job of formulating strategy has to go on to defining and evaluating the stages in which to implement the long term strategy. For this part of their job they require expert help.

The membership of top management teams is changing

Selecting appropriate strategies for change and for overseeing the management of the implementation of those strategies puts special demands on top management. Top management has to:

- Direct the integration of the information strategy with the business and organisation strategies.

- Reconcile priorities of the benefits of implementing the strategies with priorities of investing what can be very large sums of money in the IT and systems changes, and of coping with the various associated disruptions in activity and organisation.

- Manage the internal power politics of changing departmental boundaries and identities in step with the linking of activities and the integration of systems.

The tasks that have to be overseen are complex and varied. For example: technical dependencies have to be worked out (thus the planner of a direct marketing operation has to take a realistic reading of when a sound stock control and distribution system will be in place) and technical capacities and capabilities (of hardware and software and software builders) have to be respected. In addition, barriers erected by those who will lose power as a result of the changes and the redistribution of information have to be avoided or removed.

The people in the organisation have to be helped to accept and cope with the consequential changes in their work arrangements.

The responsibilities for these tasks go to the heart of an organisation and the chief executive must therefore be committed to their achievement. But the tasks require deep understanding of IT factors allied to the normal skills and knowledge of top executives. Therefore, to achieve the necessary blend of competence in the management team, a chief executive should be considering whether he needs the support in his team of a chief information officer or board member for IT. Without lessening the need for the chief executive's informed involvement, such an individual would be responsible for procuring and coordinating resources for IT. A suitable individual also needs to have the experience which allows him to understand issues of organisation direction and boundaries, power structures, and multi function activities. He has to be able to work well with the other members of corporate management in resolving the issues.

Addition of a chief information officer to the top management team is one change which may need to be considered. More importantly, to survive, top management must be motivated and able to achieve change. Responding to the challenges of IT are no different. If the responses do not come, then the question has to be asked as to how the teams approach should be changed.

IT in Midland Bank is driven by the chief executive and a board member for IT

Sir Kit McMahon, chief executive of the Midland Bank, realised that IT was critical to the Bank's future and that therefore someone with in-depth IT understanding was needed as a board member. He asked Gene Lockhart to join Midland Bank with responsibility for IT. He became a main board director and was given responsibility for group operations. Lockhart was McMahon's personal appointment. His previous career enabled him to argue through technological options with his systems people. He could discuss strategy with the planners and make sure business decisions were implemented with full use of IT. Lockhart's post gave him status equal to the heads of Midland's three principal operating divisions; UK banking, global banking and investment banking.

With the commitment of McMahon, Lockhart has had a major impact on the investments in IT and the approach of the Bank to using technology. He views the technology function like a business, a profit centre with the goal of providing effective services to the Bank's customers.

Sir Kit said when he appointed Lockhart "The effective management of information technology will be among the most important success factors for any financial institution over the years ahead. Midland will be giving increasing emphasis not only to the efficiency of its support system but also to the development of new technology led products and services".

High flyers at Morgan Stanley pass through the systems department

Investment bankers, Morgan Stanley and Company, have a policy of placing potential high flying graduate recruits into their systems department. The intention is to give them rapid merit promotions as they move up from programmer through various systems grades before going into banking work. Their aim is to achieve a computer literate management group as well as a highly motivated, high calibre systems group.

The balance of advantages between centralised and decentralised management is being shifted by developments in IT

There are several contributions of IT to this issue. Some have been discussed in Chapter 10. One important contribution comes from the development of Executive Information Systems.

Executive Information Systems (EIS) can change the balance of power between top managers and those below them, and can change the management style operating between them.

EIS is a package of software dependent on database and network technology and the capabilities of "user friendly" microcomputers. EIS is important because it gives management a capability to reach down into databases – held wherever they may be in the organisation – to create management information (for example, ratios, comparisons, exceptions) and report it immediately in easily assimilated graphics and tables on computer screens.

EIS can make it easier to get the best from decentralisation, so adding to pressures for adopting decentralised management structures.

Decentralisation has been seen as a means of securing flexibility and accountability, and of avoiding the rigidities and overhead of large central staffs. But the dangers of decentralisation have been the lessening of contact and understanding by top management at the centre, and the weakening of their control. EIS allows a top manager to have direct "dialogue" with the information held by one of the units in his organisation. It enables the top manager to reach down and past decentralised management to key indicators of performance. This facility, provided it is designed so as not to encourage top management to delve into details too far down the organisation, can improve the quality of dialogue between centre and field. This goes some way to overcoming the remoteness of the centre and to improving the ability of top management to coordinate and lead a decentralised organisation, retaining the advantages that flow from delegation and decentralisation.

With EIS assisted dialogue, top management can judge better whether it has arrangements in place in the field which make for effective and efficient management. Also, EIS affects the speed and effectiveness with which young unit managers can be developed for larger responsibilities.

EIS is changing the nature of the relationships of top managers with the managers who report to them, and the resolution of the centralisation/decentralisation issue.

IT itself presents centralisation/decentralisation difficulties

The issue of where should one place the responsibility for IT adds to the complications of the centralisation/decentralisation issue. The clothing firm Burtons chose a mixed solution. Its overall policy was to devolve full management and profit responsibility to each operating company managing director in order to maximise customer responsiveness and entrepreneurial attitudes. With respect to IT, there was a conflict to resolve. On the one hand, computer systems are a critical management resource and need to reflect management's style, and responsibility for them should therefore be with the operating companies. On the other hand, the management of IT requires expertise which can be difficult and uneconomic for a small company to support and, perhaps more important, diverts operating company management attention from its focus on customers and markets. Burton's solution was to split the traditional IT function into two. Each operating company was made responsible for its own systems analysis and planning. But a central IT function was set up to control the hardware network and provide most software tools.

In this way Burtons lessened the technical risks and diversions associated with the commodity of computer processing and left operating company management to focus on customers.

EIS helped top management to identify and resolve problems in Woolworths widely spread stores

When Richard Harker joined Woolworths as UK Store Operations Director, he needed improved information on retail performance across 1000 stores and information to keep the management team able to address exceptions and issues emerging in the weekly performance of stores while something could still be done about them.

Exploiting the availability of EIS tools, he directed the rapid implementation of a computerised information system for himself and his senior regional managers. The rapid availability of trend and exception information to all his senior retail management team helped the fast and open identification and resolution of management issues.

By further detailed involvement in the design of the sequence of "exception screen conversations", Harker was also able to use the system to help focus his management team on his general concerns and priorities.

Thus, EIS was supportive in cementing the working relationship of the senior management group, providing open and relevant information, and leading the group through priority issues.

Some questions to ask about top management arrangements

- Is top management formulating a vision for the company which will exploit the trends in IT?

- Is top management driving the investment in IT and changes in strategy?

- Does the top management team need to include a chief information officer?

- Is the top management team change-minded?

- Should the centralisation/decentralisation debate be re-examined?

PART FOUR

Gamechange

12

This book has outlined an agenda of those issues of corporate strategy and management arrangements that are likely to be generally important to review because of the effects of the IT revolution. But how should management turn the generalities into a specific plan for change?

MAKING THE CHANGES HAPPEN

Strategy formulation is the first part of change management

Strategy formulation is a management activity based on three distinct but interacting reviews:

- *Business strategy review*: an overall review of existing corporate and functional strategies in the context of changes in markets, products, supplies, competition, and technology, and in the context of changes in IT and communications capabilities such as set out in this book.

- *Organisation strategy review*: a review of management structure and roles in the context of business strategy, including changes in IT and communications capabilities.

- *Information strategy review*: a review of the information systems infrastructure required to support the business strategy and organisation, in the context of existing systems and IT and communications capabilities.

Strategy formulation thus provides targets for:

- Where should we be going?

- With what management arrangements?

- With what information systems and processing arrangements?

If it is to be useful in practice, strategy formulation must be realistic. It must pay attention to the constraints which limit the freedom of management to act: workforce culture and skills can be changed only with effort and time; the constraints of technology need time to evolve; providers of finance and customers for products have expectations which must be managed; likewise it

Strategy formulation and change: two ends of a single process

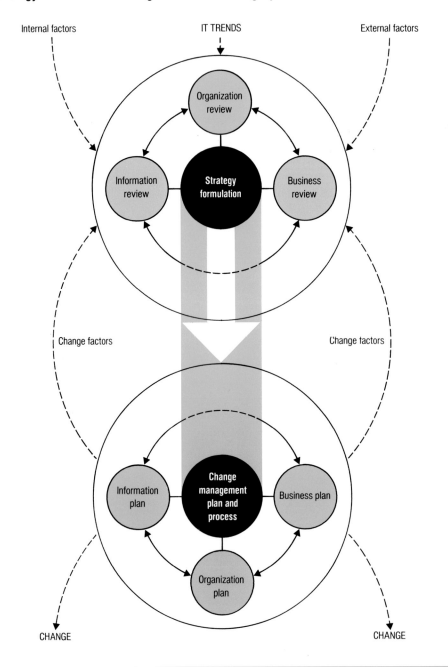

is dangerous for corporate and management security to disappoint the short-term expectations of stock markets and boards of directors. All of which is why strategy formulation ought to incorporate an intimate understanding of the pressures on management, of the possibilities open to management, and of the ambitions of management.

New strategy implies change. All change – however well planned – leads generally to surprises; some of the surprises will be unpleasant, and, sometimes, a surprise will be more important in its consequences than the problem or opportunity which provoked the change in the first place. "Beware the law of unintended consequences" should be a text for all strategy formulators. Its significance is as great for recognising and managing the consequences of new IT supported systems as for any other major corporate changes – it emphasises yet again that top management must understand and sustain the process and uncertainties of exploiting the IT revolution.

Strategy formulators should also respect the text, "if it works, don't fix it"

Do not change unless you have to – change, particularly IT induced change, will be expensive and divert management attention – but do not do it too late. One of the messages of this book is that the opportunities for change which IT creates are developing fast. This means that management need to stay constantly alert and informed if the right time for change is not to be missed. Top management need also to keep their IT enthusiasts under control if IT induced change is not to be premature. But, at the same time, top management have to appreciate the long lead times that may be necessary to design and install new IT systems and to change culture, structure, and skills so that the business can achieve the intended benefits of the systems.

Strategy formulation and change management planning and process design are two ends of a single process

The two stages of change management have to be coupled together. One without the other will be either useless or disappointing.

Strategy formulation by itself leaves unanswered:

- What do we do?

- In what order?

- How do we get people to do what has to be done?

These questions are the subject of Change Management Planning and Process Design. The first stage of this is drawing up the business plan, organisation plan, and information plan – each showing the who, what, and when for achieving the strategy, recognising the interdependencies over time. For example, it may make little sense to change management structure and roles until a sufficient amount of information infrastructure and training is in place to support the change.

The second stage is for management to draw on the three plans to build the overall change management plan and process, in which are recognised the realities of stimulating and supporting people through the difficulties of change.

Establishing both a plan for change management and a management process for leading and controlling the implementation of the plan are crucial to successful change

The plan must lay out not only the stages of development of strategy, structure and systems, but also the details of requirements for changes in amounts and timing of resources of people, facilities and finance. The resources plan must recognise the additional requirements which may be needed to take the organisation through the temporary trauma of change. And the stages of development and the resource plan must be knitted together into a schedule of activities and targets which can be used for control, just as in any other project management exercise.

Many of the changes in strategy and structure opened up by IT require nothing less than a change in the culture of the business. The changes signal discontinuities in assumptions about how to run the business and how to manage its people. The changes may require considerable investment; it is critical that they be effective. But their effectiveness depends on management successfully bringing its organisation through a number of barriers. Resistance to change has to be converted into enthusiasm for change. Motivation for change must be supported by measurement systems made to fit the new goals and arrangements. Training has to be effective. Communication has to be clear. Leadership has to be provided all the way down the organisation.

The importance of the problems has stimulated the development of a variety of techniques to assist in their solution. But above all, top management must be in charge – and that includes being in charge of IT.

Strategy formulation and change management require a mix of skills

Expertise in business, human behaviour, and information systems are all necessary. Management has to recognise and exploit the potential connections between the various types of expertise. To do this realistically, management needs access to experience about what it takes in terms of cost, time, effort and skills to implement new information systems. For sustained competitive strength, management must not underestimate the long term consequences of the newly possible information systems on business unit structure and strategy. In today's world, in the middle of the IT revolution, in the types of situation outlined in this book, this means that management and business strategists and organisation experts have to start from a realistic understanding of the problems and opportunities created by IT. And because the changes which are possible may disturb management's power structures and assumptions, help from outside the organisation may be needed to bring independence to the search for the right balance between realism and vision.

Strategy, structure, and information have to work together for corporate survival and prosperity in a competitive world. The previous chapters have aimed to highlight some of the ways in which developments in IT are changing how this may be done, and to put the issues raised on the boardroom agenda.

13

SUMMARY.
LAST THOUGHTS AND NEXT MOVES

I wanna be the leader
I wanna be the leader
Can I be the leader?
Can I? I Can?
Promise? Promise?
Yipee, I'm the leader
I'm the leader

OK what shall we do?
Roger McGough

Reprinted by permission of
Peters Fraser and Dunlop Group Ltd

IT is becoming too important to leave to IT experts

IT is shifting the ground under management. As the power and performance of IT continue to improve massively year by year, IT starts to affect the basics of corporate competitive power. IT is changing management's game.

Until now, common doctrine has been that a resolutely ambitious top management must lay the ground for sustained success in three ways. First they must do a superior job of formulating corporate strategy. Then they must define and put in place a structure and people that will facilitate implementation of the corporate strategy. Third, they should devise a practical strategy for producing the information systems to support both management of the corporate strategy and adaptation of the strategy to newly appearing opportunities and constraints.

Traditionally, strategy was seen to drive structure, and both strategy and structure were seen to drive systems. Strategy itself was driven by management's ambitions and by analysis of the company's comparative strengths and weaknesses. Now, those strengths and weaknesses can be significantly affected by what is done with IT based systems, whether by the company itself, or by competitors and potential competitors, or by suppliers and customers. So there is now also a reverse drive of systems strategy onto corporate strategy and structure.

The potentially crucial influence of systems on strategy and structure means that, in the information age, IT is becoming too important to leave to IT experts. Because the marketplace power of a corporation can be vulnerable to how well the corporation exploits the technical power of the new IT, and because creative use of IT can be the key to new corporate activity, top management must themselves define business strategy and design structure and organisation in ways that exploit the possibilities being opened up by IT.

Why must top management themselves be in charge of these tasks? Because only they have both the necessary perspective and authority. And if top management should avoid the responsibility, they can be sure that it will not be avoided by the top management of some competitor, located wherever in the world there is a potential competitive advantage waiting to be exploited, and for whom the barrier of distance is waiting to be shrunk by IT.

It is easy to say that top management should be responsible for properly exploiting the potential of IT in defining corporate strategy and structure, but the exhortation will sound unrealistic to the many top managers who are uncomfortable with IT and its experts. (With the exception, hopefully, of those managers who have read this far!)

How then can top managers cross the gulf separating them from assuming their responsibility? Their task is hard. They must avoid the trap of underestimating the long term consequences of IT – a trap which is easy to fall into if they are disappointed by experience of new IT installations whose costs and difficulties were underestimated and whose short term benefits were overestimated. The task is in two parts, one for top managers and one for top managers with the assistance of IT experts.

What top management must know and what they must do

The first part of management's task is to aspire to what can be done through IT. Not only must top management bring their understanding of the business and their imagination to their job of conceiving business strategies and structures, but they must also bring a vision enhanced by a broad understanding of what can now be achieved with IT.

If their vision is to extend to the limits of what is becoming possible, they must appreciate the reality of the underlying trend in IT capability. This trend is simply that IT is developing to the point where information which in principle is accessible to management will gradually be able to be made accessible in practice. Developments in the technology of computers and communications are shrinking the information gap which exists between the information which an organisation needs to carry out its ideal strategy in an ideal way and the information which can in practice be made available easily and economically. And as the information gap shrinks, so also does the gap between ideal strategy and management arrangements and practical strategy and arrangements. What is happening is that it is becoming practical to build systems which will give each member of an organisation's staff the power to access and analyse the data he needs. It will be practical to have access to data without delay, whether about past or current activity, and whether it has been stored in a local computer or in a computer the other side of the world. And it will be practical for different sets of data to be instantaneously combined and processed to present information in the forms which are most meaningful and convenient for staff to use in doing their jobs.

But the path to closing the information gap is likely to be difficult and expensive. How can top management recognise whether the journey must start, and in which direction, and how far should it go?

Each management will have its own way of answering the questions. One suggestion is to engage in an *open information visioning* exercise. In this step of strategy formulation, management consider opportunities for improvement which could be opened up by the potential new freedom of information. The attempt is made to conceive an ideal strategy and arrangements unfettered by practical constraints on the information which could be available in principle.

In open information visioning, each of the factors critical to the success of the business should be examined to see in what way its present achievement is bound by restraints on information, the restraints being candidates for removal by the improvements in information which are supported by the trends in IT.

In this examination, top management has to be aware of the five types of improvement identified in Chapter 4. Two of them – improving staff time economy and data accessibility – will make information easier to use, with less work; they may change the cost of information so much that they will give a competitively significant cost advantage or will make otherwise impractical activities possible. The other three – information timeliness, information computability and system adaptability – will make information more useful and may make it possible to open up new opportunities for service ahead of competitors.

Top management can then oversee production of a vision of how activities could be changed to achieve superior value and competitive power by exploiting these five types of possibilities for improving information.

The vision should include a view of the new organisation and systems required both to support the new activities and to improve corporate system performance potential through providing better information and through organising departments so that they have better focus, adaptability, accountability and economy.

Then comes the second part of top management's task – to form a view on implementation of the changes. Assessments have to be made of what would have to be actually done to make the information better, what it would cost, how long it would take, and how the change to getting and using the information would be managed. All this is likely to require knowledge and experience of IT and systems development and change, for which top management will usually have to depend on their IT experts and other advisers.

What should be top management's next moves in deciding where strategy and structure should respond to the developments in IT? One answer is first to check that there has been an imaginative, informed and up-to-date review of ways of improving the critical success factors which have emerged in this book as likely to be affected in important ways by the new power of IT. The next step for top management is to judge whether it is worth committing the investment and effort to move towards the vision of the future implied by the possibilities for improving the critical success factors. Then it is for top management and IT management together to judge how the path of corporate development should be changed in practice.

The theme of GameChange is that the judgements should be made recognising that visions of an open-information future are now worth making: visions that have been way-out are coming into reach. To decide whether vision and present reality are greatly different, and whether the vision should be striven for or left to competitors to risk, perhaps start with the questions on the next page.

Ten concluding questions for the Board

As IT changes the game, how can we go on winning? How should strategies and structures be changed to respond to the possibilities and requirements of:

- increasing product or service *variety*?

- building IT into products or services to increase the competitive edge of their *capabilities*?
- overcoming barriers of *geography*?

- exploiting potential *supply chain power*?

- leveraging the staff's *knowledge and skills*?

- facilitating *delegation and control*?

- redefining *department frontiers* to improve *economy, focus, adaptability, and accountability*?

- *recruiting, retraining, and reorientating* management and staff for new tasks and a new culture?

- *integrating* the implementation of new systems, structures and strategies?

In summary - a question to be answered by management and IT experts

■ How could we win and sustain advantage from changing *what* we do and *how* we do it by exploiting the new capabilities of IT to:

  – make it *easier and cheaper* to process, access, and use information which has been available in principle but not in practice?

  – increase the *usefulness* of various types of information by improving information timeliness and computability, and by improving system adaptability?